WHAT'S THE DEAL?

An FTC Study on Mobile Shopping Apps

August 2014 | Federal Trade Commission Staff Report

CONTENTS

EXECUTIVE SUMMARY

Through a variety of new mobile applications ("apps"), technology is rapidly changing the way consumers shop. Today's mobile apps offer new beneficial services designed to enhance the consumer shopping experience. These apps allow smartphone users to compare competing products and retailers in real-time, seek out the best deals, and pay for their purchases by waving their phones at the checkout counter. Many of these apps have been installed on millions of devices, and all use mobile technology to alter consumers' shopping.

To better understand the consumer protection implications of these emerging products and services, Federal Trade Commission staff studied some of the most popular apps that allow consumers to compare prices across retailers, collect and redeem deals, or pay for purchases while shopping in brick-and-mortar stores. Staff sought to learn more about how these apps and services operate, primarily by examining information that is available to consumers before they download the software onto their mobile devices. Staff looked for pre-download information describing how those apps that enable consumers to make purchases dealt with fraudulent or unauthorized transactions, billing errors, or other payment-related disputes. In addition, because shopping apps can allow multiple parties to gather and consolidate personal and purchase data, staff looked for information explaining how the apps handled consumer data.

Based on its review, staff found that the apps studied often failed to provide pre-download information on issues that are important to consumers. Prior to download, few of the in-store purchase apps provided any information explaining consumers' liability or describing the app's process for handling payment-related disputes. Additionally, although nearly all of the apps made strong security promises and linked to privacy policies, most privacy policies used vague language that reserved broad rights to collect, use, and share consumer data, making it difficult for readers to understand how the apps actually used consumer data or to compare the apps' data practices.

In light of these findings, staff makes the following recommendations to companies that provide mobile shopping apps to consumers:

First, when offering consumers the ability to make payments through mobile devices, **companies should disclose consumers' rights and liability limits for unauthorized, fraudulent, or erroneous transactions**. While a few of the in-store purchase apps that staff reviewed extended liability-limiting protections to consumers through pre-download representations, many provided no such disclosures. Some placed all liability for unauthorized charges on the consumer. Consumers should be able to know what their potential liability is for unauthorized transactions, what, if any, protections are available based on the method of payment, and whether procedures are available for resolving disputes, before committing to use one of these services.

Second, **companies should clearly describe how they collect, use, and share consumer data**. While almost all of the apps that staff reviewed had privacy policies purporting to address how they handle consumer data, these policies often used vague terms, reserving broad rights to collect, use, and share consumer data without explaining how the apps actually handle consumers' information. More detailed explanations would help consumers evaluate and compare the data practices of different services in order to make informed decisions about the apps they install.

Third, **companies should ensure that their strong data security *promises* translate into strong data security *practices***. Many of the surveyed apps promised to implement "technical," "organizational," or "physical" safeguards, such as data encryption, to ensure the security of consumers' data. Staff encourages all app developers (and indeed all companies in this ecosystem) to provide strong protections for the data they collect, especially in light of the technological advances found in today's smartphones that offer the potential for increased data security. And, certainly, companies must honor any commitments they make about the security they provide.

Staff also makes recommendations to consumers that use shopping apps. Specifically:

Consumers should look for the dispute resolution procedures and liability limits of the apps they download, and consider the payment methods used to fund their purchases. Federal law currently limits consumers' liability for unauthorized transactions made with their credit or debit cards, but does not limit liability for prepaid cards or accounts such as gift or general purpose reloadable ("GPR") cards. Thus, if an unauthorized charge is made or if something goes wrong with a transaction funded by a prepaid card or through an app account with a pre-funded balance, consumers might not have any recourse. If consumers cannot find information about the dispute resolution procedures and liability limits of an in-store purchase app prior to download, they should consider downloading an alternative app, or making only small-dollar purchases.

Likewise, consumers should seek information before they download apps about how their data will be collected, used, and shared. If consumers cannot find this information, or are uncomfortable with what they find, they should look for a different app or consider taking steps to minimize their exposure by limiting the personal and financial data they provide.

INTRODUCTION[1]

The growth of mobile technology, combined with the prevalence of smart mobile devices, is changing the way people shop. Today's mobile apps offer new beneficial services designed to enhance the consumer shopping experience. These apps can allow consumers to use their mobile phones in stores to compare prices against a host of competing retailers,[2] seamlessly collect and redeem coupons or discounts,[3] and pay for purchases.[4] And, while consumers and businesses explore the opportunities presented by these developments, enterprising companies and their engineers are designing tomorrow's technologies. From detailed customer tracking technologies[5] to computerized eyeglasses[6] and refrigerators that keep inventory,[7] the ways in which technology affects consumers' shopping experiences will continue to evolve.

1. The primary authors of this FTC staff survey and report are Andrew Hasty, Jared Ho, and Patti Poss of the Bureau of Consumer Protection's Mobile Technology Unit. They received valuable assistance from staff in the Division of Financial Practices, the Division of Privacy and Identity Protection, the Division of Consumer and Business Education, the Western Regional Office, the Bureau of Economics, and Rahul Murmuria, a George Mason University graduate student on temporary staff at the FTC.

2. *See, e.g.,* BOARD OF GOVERNORS OF THE FEDERAL RESERVE SYSTEM REPORT, CONSUMERS AND MOBILE FINANCIAL SERVICES 2014, at 2 (2014) (noting that 44% of smartphone owners report using their mobile phone to comparison shop while in a retail store, 68% of whom have changed where they made a purchase as a result), *available at* http://www.federalreserve.gov/econresdata/consumers-and-mobile-financial-services-report-201403.pdf.

3. *See, e.g.,* eMarketer, *Majority of US Internet Users Will Redeem Digital Coupons in 2013,* EMARKETER (Oct. 21, 2013) (finding that "more than 28% of people who own a mobile device redeemed a coupon this year"), *available at* http://www.emarketer.com/Article/Majority-of-US-Internet-Users-Will-Redeem-Digital-Coupons-2013/1010313; *see also* John Heggestuen & Marcelo Ballve, *Smartphone Coupons Are Going Mainstream, Driving Dollars and Foot Traffic to Bricks-and-Mortar Retailers,* BUSINESS INSIDER (Sep. 28, 2013) ("The number of U.S. smartphone users using mobile coupons has increased dramatically — from 7.4 million in 2010 to 29.5 million last year," predicting that 47.1 million U.S. consumers will do so in 2014), *available at* http://www.businessinsider.com/coupons-mobile-commerce-drivers-2013-9.

4. *See, e.g.,* Brad Stone & Olga Kharif, *Easy Mobile Payments Are Almost Here,* BUSINESSWEEK (Nov. 14, 2013) ("After some false starts, the next revolutionary shift in payments is gathering momentum…[w]orldwide mobile payment transactions will total $235.4 billion in 2013, a 44 percent increase from $163.1 billion in 2012, according to tracking firm Gartner"), *available at* http://www.businessweek.com/articles/2013-11-14/2014-outlook-easy-mobile-payments-in-reach.

5. *See, e.g.,* Verne Kopytoff, *Stores Sniff Out Smartphones to Follow Shoppers: Indoor location technology brings Internet-style tracking to physical spaces,* MIT TECHNOLOGY REVIEW (Nov. 12, 2013) ("Some systems use video cameras, sound waves, or even magnetic fields. In September, Apple added a feature called iBeacon to its smartphones; it emits a low-power Bluetooth radio signal, also designed for indoor use. The most widely used technique is to intercept Wi-Fi signals emitted by shoppers' smartphones. Triangulating on that signal can estimate the phone's position to within a few meters. Stores also collect a unique identifier, called a MAC address, for each phone."), *available at* http://www.technologyreview.com/news/520811/stores-sniff-out-smartphones-to-follow-shoppers/. *See also, Spring Privacy Series: Mobile Device Tracking,* FED. TRADE COMM'N (Feb. 19, 2014) (Workshop), transcript *available at* www.ftc.gov/system/files/documents/public_events/182251/140219mobiledevicetranscript.pdf.

6. *See, Internet of Things: Privacy & Security in a Connected World,* FED. TRADE COMM'N (Nov. 19, 2013) (Workshop), transcript, at 125, *available at* http://www.ftc.gov/bcp/workshops/internet-of-things/.

7. *See id.* at 127-128.

The FTC's Role in the Evolving Mobile Landscape

The Federal Trade Commission ("FTC") has extensive experience examining the opportunities and challenges associated with changing consumer technologies, including the development of mobile commerce over the last decade. Through workshops,[8] reports,[9] and law enforcement actions,[10] the FTC has worked to ensure that consumer protections keep pace as mobile commerce transitions from concept to reality. For example, at its April 2012 workshop addressing mobile payments, *Paper, Plastic…or Mobile*, FTC staff presented a snapshot look at the dispute resolution, liability limits, and data handling policies of 19 mobile payment providers,[11] and convened experts with a range of perspectives to explore the implications of existing and emerging solutions for consumer protection issues that arise in connection with mobile payments.

Following the workshop, the FTC released a report in March 2013 emphasizing the key areas for consumer concern discussed by participants at the workshop: the importance of clear dispute resolution and liability limits information, along with the need for mobile payment companies to provide greater transparency surrounding their data practices.[12] Since the release of its mobile payments report, FTC staff has continued to make emerging mobile issues a high priority — hosting additional workshops on topics like mobile cramming and security, and pursuing enforcement actions to curb unfair billing practices.[13]

8. *See Mobile Cramming: An FTC Roundtable*, FED. TRADE COMM'N (May 2013) (Workshop), *available at* http://www. ftc.gov/bcp/workshops/mobilecramming/; *Paper, Plastic…or Mobile*, FED. TRADE COMM'N (April 2012) (Workshop), *available at* http://www.ftc.gov/bcp/workshops/mobilepayments/; *Pay on the Go: Consumers and Contactless Payment*, FED. TRADE COMM'N (July 2008) (Workshop), transcript *available at* http://www.ftc.gov/bcp/workshops/ payonthego/index.shtml; *Protecting Consumers in the Next Tech-ade*, FED. TRADE COMM'N (2006) (Workshop), *available at* http://www.ftc.gov/bcp/workshops/techade/what.html; *The Mobile Wireless Web, Data Services and Beyond: Emerging Technologies and Consumer Issues*, FED. TRADE COMM'N (Feb. 2002) (Workshop), *available at* http://www.ftc.gov/bcp/workshops/wireless/index.shtml.

9. FED. TRADE COMM'N, PAPER, PLASTIC…OR MOBILE (2013), *available at* http://www.ftc.gov/sites/default/files/ documents/reports/paper-plastic-or-mobile-ftc-workshop-mobile-payments/p0124908_mobile_payments_ workshop_report_02-28-13.pdf; FED. TRADE COMM'N, RADIO FREQUENCY IDENTIFICATION: APPLICATIONS AND IMPLICATIONS FOR CONSUMERS (2005), *available at* http://www.ftc.gov/sites/default/files/documents/reports/rfid-radio-frequency-identification-applications-and-implications-consumers-workshop-report-staff/050308rfidrpt.pdf.

10. *Credit Karma, Inc.*, FTC File No. 1323091 (F.T.C. Mar. 28, 2014), *available at* http://www.ftc.gov/enforcement/ cases-proceedings/132-3091/credit-karma-inc; *Fandango, LLC*, FTC File No. 1323089 (F.T.C. Mar. 28, 2014), *available at* http://www.ftc.gov/enforcement/cases-proceedings/132-3089/fandango-llc; *FTC v. Wise Media, LLC*, No. 1:13-cv-1234 (N.D.Ga. 2013), *available at* http://www.ftc.gov/enforcement/cases-proceedings/122-3182/ wise-media-llc-et-al; *FTC v. Tatto, Inc.*, No. 2:13-cv-08912 (C.D.Ca. June 13, 2014), *available at* http://www. ftc.gov/enforcement/cases-proceedings/1123181/tatto-inc-also-dba-winbigbidlow-tatto-media-et-al; *FTC v. Jesta Digital, LLC*, No. 1:13-cv-01272 (D.C. Aug. 21, 2013), *available at* http://www.ftc.gov/enforcement/cases-proceedings/112-3187/jesta-digital-llc-also-dba-jamster; *Apple, Inc.*, Docket No. C-4444 (Mar. 25, 2014), *available at* http://www.ftc.gov/enforcement/cases-proceedings/112-3108/apple-inc.

11. *See A Snapshot of Select Mobile Payment Providers' Disclosures: FTC Staff's Preliminary Observations*, FED. TRADE COMM'N (Apr. 2012) (Workshop), *available at* http://www.ftc.gov/bcp/workshops/mobilepayments/.

12. PAPER, PLASTIC…OR MOBILE, *supra* note 9 at 5-7, 11-15.

13. A list of FTC mobile technology initiatives and enforcement matters can be found at http://www.ftc.gov/opa/ reporter/mobile/index.shtml.

The 2012 workshop and subsequent report examined the disclosures and dispute resolution processes for mobile payment apps. To expand on this work, FTC staff designed a survey focusing on apps that consumers use to shop and make purchases. Specifically, staff reviewed app promotion pages in Google Play and the iTunes App Store to find three categories of apps: (1) apps that facilitate real-time price comparison (*i.e.*, "price comparison apps"); (2) apps that enable users to find and redeem coupons or discounts (*i.e.*, "deal apps"); and (3) apps that allow consumers to make purchases in physical stores (*i.e.*, "in-store purchase apps"). Staff then closely examined the app promotion pages, developer websites, and other pre-download information for the 25 most installed apps for each of the three functionalities, in each app store.[14]

In addition to exploring the apps' mechanics and features, staff looked for information about how the apps handle payment disputes and consumer data. For this information staff turned to any license agreement, privacy policy, terms document, or other developer-provided disclosure that appeared to be associated with each app prior to download. With respect to the in-store purchase apps, staff examined what the apps' disclosures said about fraudulent or unauthorized transactions, billing errors, or other payment-related disputes. Because all three categories of apps can collect, use, and share personal data, staff also examined disclosures regarding their data-handling procedures. That is, before a consumer decides to install one of these apps, what can consumers learn about the app's data practices?[15]

In general, staff encountered a wide range of innovative services that used mobile-specific features to present consumers with new ways of researching products, finding deals, and paying for their purchases. However, staff also found that it was often difficult to determine, based on pre-download information, what protections were available to users of in-store purchase apps in case of payment-related disputes. Additionally, while many apps in all three categories made strong security promises and linked to privacy policies, most privacy policies used vague language reserving broad rights to collect, use, and share consumer data, making it difficult for readers to understand how the apps actually handled consumer data.

Based on these findings, staff reiterates its call for those operating these services to publish clear dispute resolution and liability limits information and to provide greater transparency regarding their data-handling

---

14. In order to estimate each app's relative number of installs, staff relied on the download ranges provided by Google Play and the total number of feedback ratings provided by the iTunes App Store. *See* Appendix A at A-3.

15. While staff's review centered on the pre-download information available to consumers, staff also downloaded and opened each app to ensure that the information reviewed and analyzed was as accurate. Staff did not sign up for any accounts. It is possible that the apps provide their users with additional disclosures and information as the user interacts with the service. *See* Appendix A at A-4 - A-5.

practices.[16] Staff also encourages consumers to look for and review information about how these services work, what they can do if they encounter a problem, and how their data is collected, used, and shared. If consumers cannot find this information, they should consider taking steps to minimize their risks by limiting the personal and financial data they provide, or by choosing a different app.

16. In addition to providing greater transparency, staff also reiterates its call for companies to practice "Privacy by Design" and to offer consumers simplified choices over how their data is handled. *See* FED. TRADE COMM'N, PROTECTING CONSUMER PRIVACY IN AN ERA OF RAPID CHANGE, RECOMMENDATIONS FOR BUSINESSES AND POLICYMAKERS (2012), *available at* http://ftc.gov/os/2012/03/120326privacyreport.pdf (Commissioners Ohlhausen and Wright were not members of the Commission at the time that the Privacy Report was issued and thus did not offer any opinion on that matter); *see also* FED. TRADE COMM'N STAFF, MOBILE PRIVACY DISCLOSURES: BUILDING TRUST THROUGH TRANSPARENCY (2013), *available at* http://www.ftc.gov/sites/default/files/documents/reports/mobile-privacy-disclosures-building-trust-through-transparency-federal-trade-commission-staff-report/130201mobileprivacyreport.pdf.

SURVEY OVERVIEW

Google Play and the iTunes App Store currently offer a number of apps designed to enhance consumers' shopping experiences. Although consumers have not yet abandoned their traditional wallets, staff found a diverse pool of apps enabling consumers to compare in-store prices and to find and redeem deals at the point of sale. As detailed further in Appendix A, FTC staff examined thousands of app promotion pages in Google Play and the iTunes App Store to find price comparison apps, deal apps, and in-store purchase apps. Staff then reviewed the pre-download information associated with the 25 most installed apps in each store for each of the three functionalities of interest. After removing apps that were not available for US consumers to download, staff reviewed a total of 121 unique apps — 60 Google Play apps and 61 iTunes apps. The reviewed apps included 47 price comparison apps, 50 deal apps, and 45 in-store purchase apps (a number of these apps appeared in more than one of the three categories).[17]

All of the apps staff reviewed had two traits in common: popularity and low install costs. Thirty-seven (62%) of the 60 total Google Play apps that staff reviewed had been downloaded more than 1 million times, and 18 (30%) had been downloaded more than 5 million times. While download information for iTunes apps is not available, the number of user ratings associated with the iTunes apps suggests a significant number of installs. Specifically, 37 (61%) of the 61 total iTunes apps had received more than 5,000 user ratings, and 13 (21%) had more than 45,000. In terms of cost, 115 of the 121 total apps (95%) were offered free of charge, and the remaining six cost less than two dollars to download.

Price Comparison Apps

Since many consumers report using their smartphones to comparison shop,[18] staff reviewed the pre-download information for the 47 apps designed to allow users to compare prices, in real-time, while shopping in brick-and-mortar stores. From apps that helped consumers locate the cheapest gasoline in their neighborhoods to apps that scanned Universal Product Codes ("UPC codes") and displayed product prices at local and online retailers, staff encountered a wide range of price comparison apps.

Nineteen of the price comparison apps expressly promoted their ability to compare product prices in order to help users find the best bargains, and virtually all (43) of the price comparison apps promised to provide users with access to other information, such as product details and availability, user reviews, and

---

17. Three price comparison apps and five in-store purchase apps were removed from the survey because they did not have a US presence or were no longer available for download. While staff restricted the scope of its review to apps with a presence in the United States, mobile shopping and payment apps are developing throughout the world. *See* Organisation for Economic Co-operation and Development (OECD), Consumer Policy Guidance on Mobile and Online Payments (2014), *available at* http://www.oecd.org/sti/consumer/consumer-policy-guidance-on-mobile-and-online-payments.htm (offering policy guidance to strengthen consumer protections surrounding the rapid growth of e-commerce and mobile payments in an international context).

18. *See* Consumers and Mobile Financial Services 2014, *supra* note 2.

merchant locations.[19] Some of the apps (16) promoted a connection with social media, often allowing users to share product information via specific social media networks. Lastly, some (16) promised to provide results related to a user's location by showing products' prices at nearby businesses.

Most (38) of the apps appeared to work by using the mobile device's camera to scan product codes or take pictures of the products themselves. The apps would then use image recognition technology to identify the specific barcode or product in the users' photos, which enabled them to search various databases and return prices and other information from different online and brick-and-mortar retailers.[20]

Deal Apps

Staff also looked at apps that promised to enable smartphone users to find and redeem discounts in physical stores. Staff reviewed a total of 50 deal apps, which generally fell into one of two categories. Seventeen apps were offered by a specific company for their customers to use in brick-and-mortar stores, while 26 apps aggregated coupons and discounts from a number of different stores, restaurants, and service providers. The remaining seven apps included two from financial companies (like credit card providers), which promised their customers retailer-specific specials; four from "deal of the day" discount services, promising to offer discounts on various products and services each day; and one from a merchant whose deals were redeemable only through its online storefront.

As with the price comparison apps, the deal apps that staff reviewed often used mobile devices' unique features to promise consumer savings. Forty-eight of the deal apps indicated that a user's physical location could affect the coupons or discounts offered. Some of these apps (18) appeared to obtain a user's location manually (*e.g.*, by requesting that a user enter a zip code or postal address), while most (45) appeared to obtain a user's location automatically (e.g., via GPS, Wi-Fi or cell tower information, or some other automated detection technique). Many (26) of the apps using location information displayed "weekly sale" ads, coupons, or other discounts from retailers that were within the user's vicinity, and a few (15) indicated they would send discounts automatically to their users through "push notifications" or text messages based on the user's proximity to a relevant business. Taking advantage of other characteristics unique to mobile devices, a number of apps (16) suggested that consumers could accrue discounts through various activities or by performing tasks, such as walking into a store, scanning particular items, or buying certain products.[21]

---

19. Nine of the apps in this category were offered by specific retailers for their customers to use while in the retailer's store in order to obtain price and additional product information. Presumably, these apps could be used by consumers in competitors' stores to compare competitors' prices, or to see how a retailer's online prices match its in-store prices.

20. Besides UPC codes, many of these apps advertised the ability to return information associated with Quick Response ("QR") codes, European Article Number ("EAN") codes, and International Standard Book Number ("ISBN") codes.

21. Many of these apps also integrated with merchants' loyalty programs, offering coupons or discounts based on accumulated points.

Finally, most (48) deal apps touted the fact that consumers did not need to bring paper coupons with them to redeem their deals. Many of these apps allowed consumers to redeem their discounts by displaying a particular deal on their phone's screen at a merchant's location, but some worked by associating a user's deals with the user's merchant-specific loyalty cards. Specifically, 26 followed a growing trend that allows consumers to use their smartphones to connect their digital coupons with their loyalty cards so that consumers' deals can be redeemed automatically whenever they use their loyalty card at a checkout terminal.[22]

In-Store Purchase Apps

Apps that allowed consumers to use their phones to make in-store purchases were the hardest of the three functionalities to find. Despite reviewing thousands of app promotion pages, staff found only 30 apps that clearly enabled consumers to make in-store purchases using their smartphone.[23] These 30 apps were diverse in scope, allowing consumers to make real-world purchases via different transaction technologies, funding sources, and payment methods.

Near Field Communications ("NFC") technology — which allows two devices, such as a smartphone and a checkout terminal, to communicate through short range radio waves to effect a payment — has received a lot of attention in discussions related to mobile payments.[24] While most smartphones currently in use do not have the technology, the majority of smartphone manufacturers have begun to include NFC in their handsets, and analysts predict a significant increase in the prevalence of NFC-enabled devices over the next five years.[25] In this survey, only two of the in-store purchase apps that staff reviewed appeared to use NFC to allow consumers to make purchases.

Instead, most of the in-store purchase apps that staff encountered relied on barcode scanning technologies to process transactions, requiring the user or merchant to scan a unique code — like a UPC or QR code — at the point of sale. Seventeen of the in-store purchase apps functioned by assigning consumers a code that merchants could scan from the phone's screen in-store at the checkout terminal, while four apps assigned a code to each merchant that consumers could scan at the checkout terminal. Other apps required

---

22. *See, e.g.*, Kristen Cloud, *Report: Coupon Redemption Remains Strong, Digital Gains in Popularity*, Theshelbyreport. com (Jan. 17, 2014) (noting that "coupons that consumers load directly to their shopper loyalty accounts" and which are "applied automatically at checkout…when the shopper presents their loyalty card or unique individual identifier" have "continued to grow ahead of the overall rate of coupon growth."), *available at* http://www. theshelbyreport.com/2014/01/17/report-coupon-redemption-remains-strong-digital-gains-in-popularity/.

23. As discussed further in Appendix A at A-4, staff initially found 45 apps that appeared to contain an in-store purchase functionality. However, on closer review, seven of these apps appeared to lack the ability to make in-store purchases while eight did not provide enough detail to confirm that they contained such functionality.

24. *See* PAPER, PLASTIC …OR MOBILE, *supra* note 9, FN18.

25. *See, e.g.*, Press Release, IHS, NFC-Enabled Cellphone Shipments to Soar Fourfold in Next Five Years, (Feb. 27, 2014) (noting that "[t]he majority of smartphone makers are adopting the NFC wireless communications and payment technology in their products as a de facto standard"), *available at* http://press.ihs.com/press-release/design-supply-chain/nfc-enabled-cellphone-shipments-soar-fourfold-next-five-years.

users to type codes into retailers' terminals, and staff found one app that confirmed transactions by matching users' faces with their pictures.[26]

---

26. This service required users to associate their picture with their account, such that when a user entered a participating store the user's picture appeared on the store's computers — enabling the user to check out simply by appearing at the point of sale and affirming the transaction to a clerk.

ANALYSIS

The shopping apps reviewed by staff allow consumers to compare prices, find and redeem deals, and make in-store purchases in ways not possible just a few years ago. At the same time, they contain features that could lead to payment disputes, and raise questions about how the data they collect will be stored, used, and shared. To determine what information consumers receive about both of these important issues, staff reviewed (1) the in-store purchase apps' pre-download disclosures about consumers' liability or redress rights in case something goes wrong with a purchase, and (2) all of the apps' pre-download disclosures about what data the apps collect, how it will be used, with whom it will be shared, and how it will be secured. Staff's findings are detailed below.

Liability Limits and Dispute Resolution

As detailed in staff's Mobile Payments Report, federal laws limit consumers' liability for unauthorized payments in many contexts, but coverage can vary based on how a consumer's purchase is funded and processed.[27] In particular, when a consumer makes a purchase on an app by placing a charge directly on a credit or debit card, the consumer is protected by the same federal liability limits applicable to physical credit or debit cards. However, these statutory protections generally do not apply to purchases involving prepaid, gift, or stored value accounts. Instead, for these transactions, consumers must rely on whatever protections, if any, are voluntarily provided.

As part of its review, staff first looked for language describing how purchases were funded and processed along with information about consumers' dispute resolution procedures and liability limits.

Staff found that 14 of the 30 in-store purchase apps did not disclose whether they had any dispute resolution or liability limits policies prior to download. Further, of the 16 apps that provided pre-download information about dispute resolution procedures or liability limits, only nine offered any written protections for their users. Seven app providers disclaimed all liability through language such as "[App Provider] shall not be responsible for any losses arising from the financial loss or theft of User Information due to unauthorized or fraudulent transactions related to the Application."

Staff then evaluated what protections were likely to apply to these apps, based on the apps' different payment models and funding sources, and how these protections match the disclosures provided.

---

27. *See* PAPER, PLASTIC... OR MOBILE, *supra* note 9, at 5-7.

A NOTE ON STATUTORY PROTECTIONS

When a consumer makes a purchase using a physical payment card, the consumer's statutory protections track the type of card used. While federal law limits consumers' liability for credit and debit card transactions, and provides dispute resolution procedures for errors,[28] prepaid and gift card users must generally rely on their contracts with the card providers for these protections.[29]

When a consumer makes a purchase through an app by placing a charge directly on a credit, debit, or prepaid card — a "pass through" payment model — the consumer has the same statutory and contractual protections as if the consumer had used a physical card.

However, some mobile payment services require a consumer to move money from a traditional funding source — such as a credit, debit, or prepaid card — into a stored value account maintained by the app provider and used to pay for the consumer's subsequent purchases — a "stored value" payment model. Since these stored value services do not pass charges directly through users' funding sources, their users may not have the same statutory and contractual protections that exist in the context of conventional transactions.

---

28. Liability for unauthorized credit card charges is limited by law to $50. 12 C.F.R. 1026.12. Liability for unauthorized debit card charges is limited to $50 if reported within two business days after discovery, or up to $500 if reported after two business days. 12 C.F.R. 1005.6. Consumers that fail to report unauthorized debit card transactions within 60 days (from the date the statement was transmitted to the consumer that first shows the error) may have unlimited liability. *Id.* Federal regulations also establish dispute resolution procedures for errors associated with credit cards, *see* 12 C.F.R. 1026.13, and debit cards, *see* 12 C.F.R. 1005.11.

29. Federal law provides some liability limits for unauthorized transactions, and some procedures for dispute resolution, for payroll card and electronic benefits transfer card transactions. *See*, Comments of the Staff of the FTC Bureau of Consumer Protection before the Consumer Financial Protection Bureau ("CFPB"), *In the Matter of Request for Comment on Advance Notice of Proposed Rulemaking Electronic Fund Transfers (Regulation E) and General Purpose Reloadable Prepaid Cards*, Docket No. CFPB-2012-0019 (July 23, 2012), at 4, *available at* http://www.ftc.gov/os/2012/07/120730cfpbstaffcomment.pdf.

PASS THROUGH MODELS

Twenty-two of the 30 apps followed a "pass-through" payments model, allowing consumers to pass charges through the app directly to an external funding source like a credit, debit, or prepaid card.[30] Users of these 22 apps therefore received the same statutory and contractual protections associated with the external funding sources used to pay for the purchases. Twelve of these pass-through apps were silent regarding their dispute resolution and liability limits policies, and four expressly disclaimed all liability in their pre-download information. This means that consumers may not know they have the same statutory and contractual protections associated with these external funding sources.

STORED VALUE MODELS

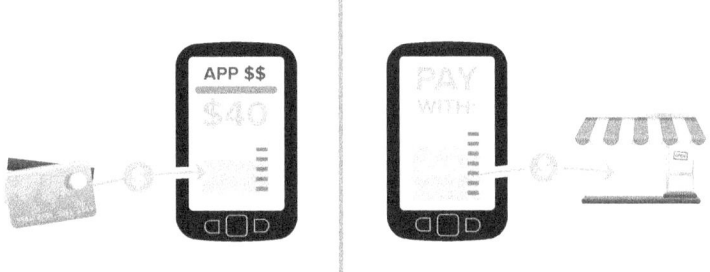

The remaining eight apps followed a "stored value" model — requiring users to move money from an external funding source into an account from which the user's charges could be deducted. Since these stored value services do not pass charges directly through to users' funding sources, their users may not have the statutory and contractual protections that exist in the context of a pass-through purchase. Two of these stored value apps were silent regarding their dispute resolution and liability limits policies, and three expressly disclaimed all liability, in their pre-download information. Only three provided policies that offered consumers any protections. This means that most users of the reviewed stored-value services would not be able to rely on any statutory or written contractual protections in the event of an unauthorized transaction or an error.[31]

30. While most of the pass-through apps allowed consumers to fund their transactions with credit or debit cards, at least fifteen also allowed consumers to use gift or prepaid cards as their funding sources, three allowed consumers to use ACH bank drafts, and one app allowed consumers to use Bitcoin transfers.

31. The 2012 mobile payments report recognizes that companies may voluntarily provide additional protections but that such protections "are not consistent, and companies that provide them could withdraw or modify them at their discretion." PAPER, PLASTIC…OR MOBILE, *supra* note 9 at 7.

Did the apps offer any protections for payment related disputes?

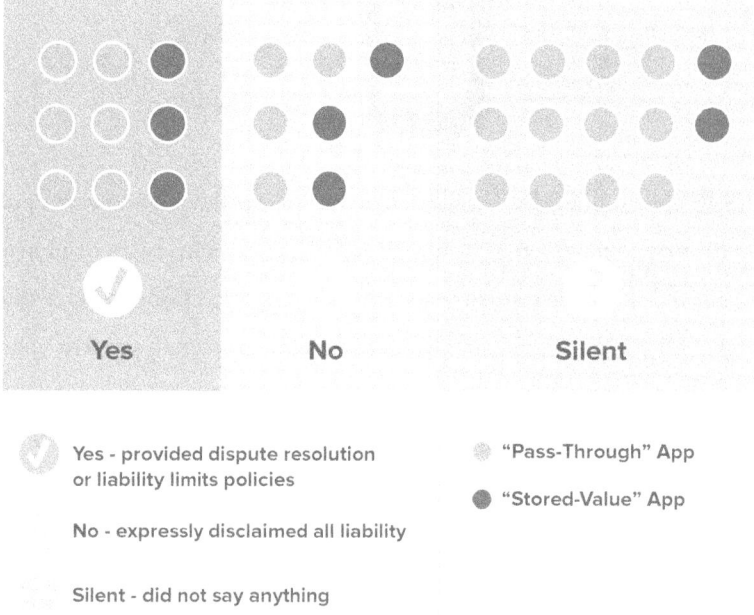

Yes   No   Silent

 Yes - provided dispute resolution   "Pass-Through" App
 or liability limits policies   "Stored-Value" App

No - expressly disclaimed all liability

Silent - did not say anything

Further complicating matters for consumers is the difficulty in distinguishing between pass-through and stored value apps in order to determine whether they can rely on any statutory or contractual protections. Staff often had to look through pages of terms and conditions, license agreements, and developer-provided "help" or "frequently asked questions" websites in order to make such a distinction — if, in fact, staff was even able to make such a distinction based on pre-download information.[32] Since consumers may not always be able to tell the difference between the two transaction models, and because consumers may not be aware that their statutory protections depend on the underlying funding source, consumers using funding sources with statutory protections (like credit or debit cards) may not realize the importance of a particular service's contractual protections.

RECOMMENDATIONS

Consumers should be able to learn about their rights and protections before using a service to make a payment via their mobile device. As described above, in-store purchase apps process transactions in ways that implicate consumers' rights and liability limits. Many transactions on mobile devices — notably, those made using stored value accounts — are generally not protected by the statutory and/or contractual liability limits that consumers might expect based on their traditional shopping experiences. Therefore, staff reiterates

---

32. As previously noted, and as described in the Appendix at A-4, staff encountered eight apps that did not provide staff with enough pre-download information to confirm that the app allowed consumers to make in-store purchases, let alone how payments were funded and processed.

its recommendation from the Mobile Payments Report that companies provide clear dispute resolution and liability limits information to their customers, particularly when using a stored value method to process payments.[33]

In addition, given the lack of alternative remedial avenues for stored-value service users,[34] consumers should look for services that tell them upfront how the payment service works and what they can do if they encounter a problem. If this information is not available, consumers should consider taking steps to minimize their liability by choosing a different payment app or funding such payments with low-dollar amounts.

Privacy and Security

Staff also looked at the privacy policies for all three categories of mobile shopping apps. More than other types of technology, mobile devices are typically personal to an individual, always on, and always with the user.[35] These characteristics enable apps to collect a significant amount of user data — such as one's location, interests, and affiliations — and share this data with a host of third parties.[36] In this way, mobile shopping apps can allow different retailers, advertisers, and other third parties to gather and consolidate an app user's personal and purchase data.[37]

For the apps in the survey, staff looked at the extent to which the policies explained what consumer data would be collected, how the data would be used, with whom the data would be shared, and how the data

---

33. *See* PAPER, PLASTIC… OR MOBILE, *supra* note 9, at 5-7. *See also* COMMENTS OF FTC STAFF, *supra* note 31, at 6-7 (In staff's comment to the CFPB regarding its general purpose reloadable (GPR) card rule-making process, staff identified the scenario where a merchant inadvertently debits a GPR card multiple times. Such errors, if not corrected, clearly cause consumer injury. Without an adequate dispute resolution process to resolve errors and liability limits for unauthorized transactions, consumers using GPR cards to fund payment apps are left with uncertainty or little to no recourse.)

34. The CFPB has issued an Advance Notice of Proposed Rulemaking (ANPR) for Electronic Fund Transfers (Regulation E). 77 Fed. Reg. 30923 (May 24, 2012). The ANPR requested comment on "open loop" GPR cards and devices. Open loop prepaid cards and devices allow consumers to make purchases "anywhere that accepts payment from a retail electronic payments network, such as Visa, MasterCard, American Express or Discover." In contrast, closed loop prepaid cards, such as gift cards, may only be used at specific merchants or merchant groups. The CFPB requested comment on how Regulation E protections should be extended to GPR cards and devices. Staff notes that even if the CFPB were to extend Regulation E protections to GPR cards, after issuing an NPRM, it is not clear if it would fully cover consumers using all types of prepaid cards or stored value items, including, for example, closed loop prepaid gift cards or close looped store-value apps. To the extent that any such items are not ultimately covered, these consumers still may only receive the contractual protections, if any, provided in their giftcard or app contracts.

35. *See* MOBILE PRIVACY DISCLOSURES, *supra* note 16, at 2-3.

36. *See id.*

37. *See, e.g.,* PAPER, PLASTIC… OR MOBILE, *supra* note 9, at 13 (noting that "[i]n addition to the banks, merchants, and payment card networks present in traditional payments systems, mobile payments often involve new actors such as . . . application developers, and coupon and loyalty program administrators" who "may have access to more detailed data about a consumer and the consumer's purchasing habits as compared to data collected when making a traditional payment").

would be secured. While most apps made strong security promises and linked to privacy policies, staff found that these policies often used vague terms, reserving broad rights to collect, use, and share consumer data and making it difficult for readers to understand how the apps actually used consumers' information or to compare the apps' data practices.

AVAILABILITY AND SCOPE OF THE APPS' PRIVACY POLICIES[38]

An overwhelming majority of the surveyed apps had privacy policies. Many could be accessed via links on the apps' promotion pages in the iTunes or Google Play app stores, and most could be found on the developers' websites.

Chart 1: Apps Linking to Privacy Policies

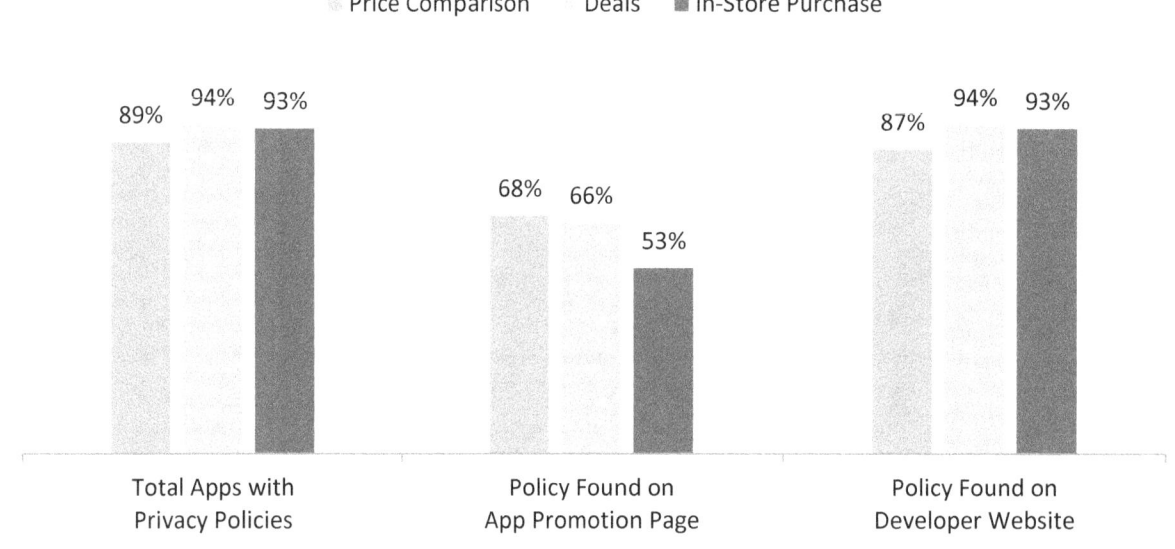

Sample Size: Price Comparison: n = 47 | Deals: n = 50 | In-Store Purchase: n = 45

Most of the privacy policies, whether found on the app promotion pages or the developers' websites, expressly covered the companies' mobile apps within the scope of the privacy policies. The rest of the companies' privacy policies implicitly covered mobile apps by stating that the policy covered the "Services" before describing practices directly related to a company's mobile app. No privacy policies excluded the companies' apps from their scope.

---

38. As discussed further in Appendix A at A-4, three of the price comparison apps and five of the in-store purchase apps did not appear to be available for consumers in the United States, leaving staff with 47 price comparison apps, 50 deal apps, and 45 in-store purchase apps. Because several apps fell within more than one category, staff reviewed the privacy policies associated with 121 unique apps.

Chart 2: The Privacy Policies' Scope

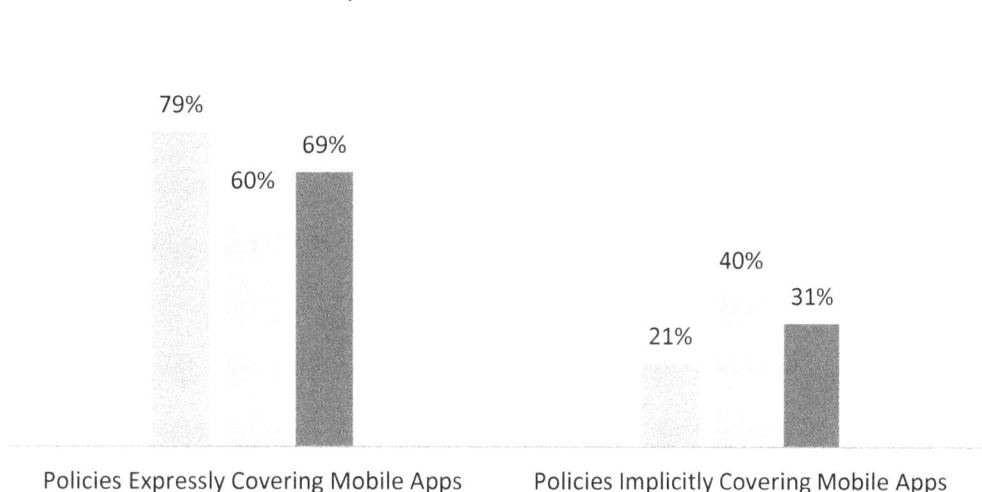

Price Comparison   Deals   ■ In-Store Purchase

79%
60%
69%

40%
21%
31%

Policies Expressly Covering Mobile Apps   Policies Implicitly Covering Mobile Apps

Sample Size: Price Comparison: n = 42 | Deals: n = 47 | In-Store Purchase: n = 42

Transparency has long been a core component of the FTC's privacy initiatives,[39] and these findings demonstrate that progress has been made in recent years.[40] However, while making a privacy policy or

39. *See, e.g.,* PROTECTING CONSUMER PRIVACY IN AN ERA OF RAPID CHANGE, *supra* note 16, at 60-64; *see also* FED. TRADE COMM'N, PRELIMINARY FTC STAFF REPORT: PROTECTING CONSUMER PRIVACY IN AN ERA OF RAPID CHANGE: A PROPOSED FRAMEWORK FOR BUSINESSES AND POLICY MAKERS, at 6-9 (2010), *available at* http://www.ftc.gov/sites/default/files/documents/reports/federal-trade-commission-bureau-consumer-protection-preliminary-ftc-staff-report-protecting-consumer/101201privacyreport.pdf.

40. Indeed, in previous surveys, FTC staff has reported finding very little information about apps' data handling policies. *See* FED. TRADE COMM'N, MOBILE APPS FOR KIDS: DISCLOSURES STILL NOT MAKING THE GRADE, at 4 (Dec. 2012) ("most [kids] apps failed to provide *any* information about the data collected through the app"), *available at* http://www.ftc.gov/sites/default/files/documents/reports/mobile-apps-kids-disclosures-still-not-making-grade/121210mobilekidsappreport.pdf; *see also* FED. TRADE COMM'N STAFF, MOBILE PrivacyDisclosures, *supra* note 16, at 7, 22 (noting a lack of privacy disclosures as revealed by staff's Kids App Reports and providing that "apps should have a privacy policy and make that policy available through the platform's app store"). Although progress has been made, staff notes that a significant percentage of apps surveyed still did not include a link to their privacy policies in the app stores.

disclosure available to consumers is a step towards increased transparency, the goal of transparency is to enable consumers to learn how, and for what purposes, companies collect, use, and share their data.[41]

THE CONTENT OF THE PRIVACY POLICIES

Staff's review of the surveyed apps' privacy policies centered on three questions: what data would be collected, how would it be used, and with whom would it be shared.[42]

DATA COLLECTION

Almost every privacy policy had a section describing the data that an app might collect. Nearly all of the privacy policies revealed that the apps might collect users' names and contact information, such as postal or email address, or telephone number.

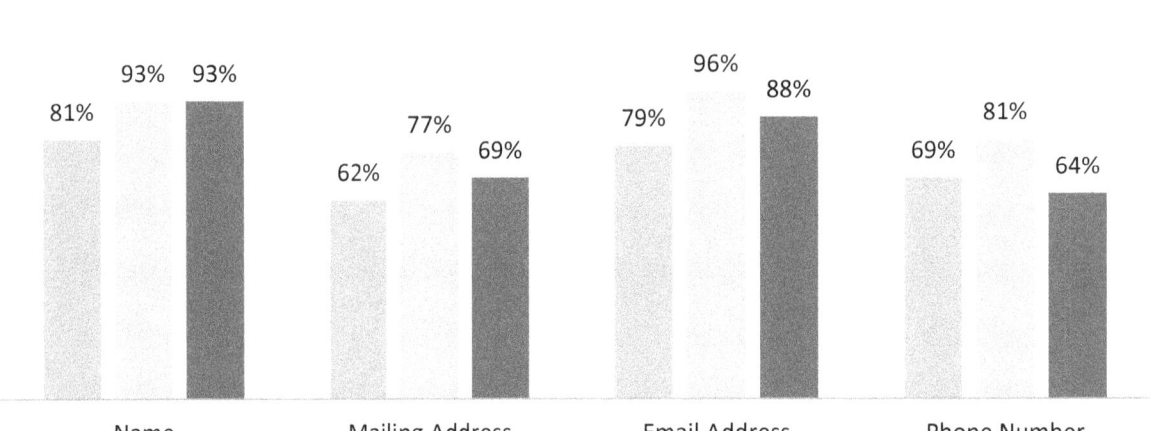

Chart 3: Collection of Name and Contact Information

Price Comparison   Deals   In-Store Purchase

Sample Size: Price Comparison: n = 42 | Deals: n = 47 | In-Store Purchase: n = 42

---

41. *See* Fed. Trade Comm'n, Protecting Consumer Privacy in an era of rapid change, *supra* note 16, at 60-64. This report is consistent with the Commission's view that while improved privacy policies are necessary, there are other methods for effectively providing this information to users during the course of business. While staff focused on the content of the apps' privacy disclosures and the information available to consumers making the decision to download these shopping apps, other aspects — such as the method by which privacy disclosures are relayed to consumers — are also very important. In particular, companies should provide just-in-time notice and choice where feasible. *See* Fed. Trade Comm'n Staff, Mobile Privacy Disclosures, *supra* note 16, at ii-iii (calling for app developers to "[p]rovide just-in-time disclosures and obtain affirmative express consent before collecting and sharing sensitive information").

42. *See* Appendix A at A-3 for a more detailed explanation of how staff conducted its review.

In addition to collecting name and contact information, many apps' privacy policies revealed that other personal data — such as a user's Social Security number, driver's license, location, date of birth, or gender — also might be collected.

Chart 4: Collection of Other Personal Information

Price Comparison   Deals   ■ In-Store Purchase

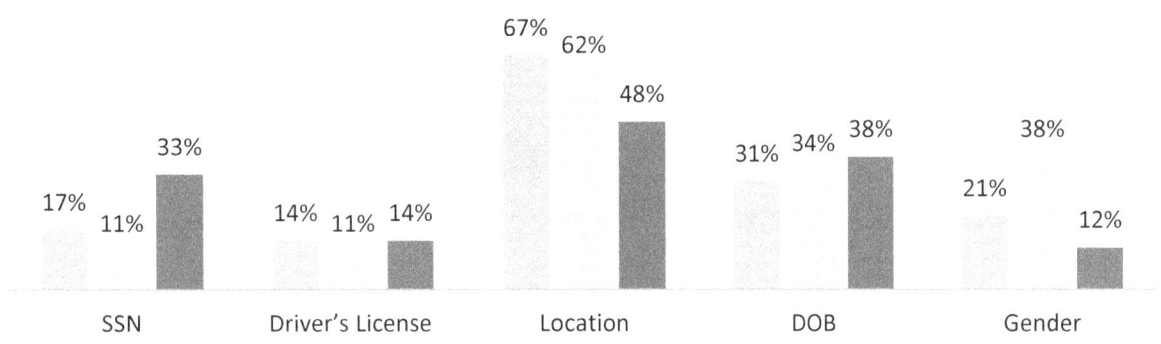

Sample Size: Price Comparison: n = 42 | Deals: n = 47 | In-Store Purchase: n = 42

Besides personal data, the privacy policies often stated that the services might collect transaction data, such as purchase details, where possible.

Chart 5: Collection of Transaction Data

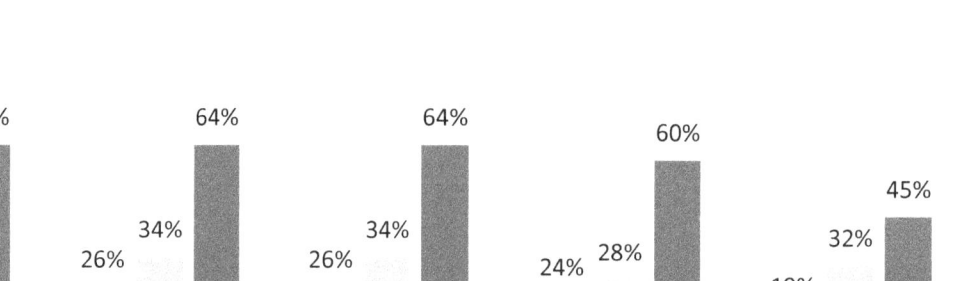

Sample Size: Price Comparison: n = 42 | Deals: n = 47 | In-Store Purchase: n = 42

Finally, a number of policies revealed that the services might obtain consumer data from third parties, including credit bureaus and identity verification services.

Chart 6: Collection of Consumer Data
From a Third Party

Sample Size: Price Comparison: n = 42 | Deals: n = 47 | In-Store Purchase: n = 42

DATA USE

Nearly all of the privacy policies that staff reviewed contained broad and vague statements regarding how the services used data, making it difficult to assess how the data would actually be used.[43] For example, some policies stated that they might use personal data to "enhance" or "improve" users' shopping experiences without further explaining those terms or providing examples that would help consumers understand the reasonable limits of such use or how uses might go beyond what consumers reasonably expect.[44] Other policies described the apps' "primary purpose" for collecting data without describing or disclaiming any potential secondary purposes. Despite frequent use of this vague language, staff identified several trends across the different apps.

Every privacy policy stated that the collected data was necessary to provide the service, and most specified additional potential uses. Of the additional potential uses, "advertising" or "marketing" was the most common, followed by "communications" with the user, "service improvements," and "personalization." Seven of the policies specifically provided that collected data could be used for reporting to credit bureaus, but such detailed disclosure tended to be the exception rather than the rule.[45]

---

43. *See* Fed. Trade Comm'n, Protecting Consumer Privacy in an era of rapid change, *supra* note 16, at 27 (providing that "[g]eneral statements in privacy policies . . . are not an appropriate tool to ensure [a reasonable limit on the collection of consumer data] because companies have an incentive to make vague promises that would permit them to do virtually anything with consumer data").

44. *See id.* (recognizing the need for flexibility to permit innovative new uses of data that benefit consumers but at the same time recognizing some reasonable limits.)

45. To the extent shopping apps may share this type of data with data brokers, staff notes the Commission's recent recommendation that Congress consider requiring consumer-facing entities to provide a prominent notice to consumers that they share consumer data with data brokers and provide consumers with choices about the use of their data, such as the ability to opt-out of sharing their information with data brokers. Fed. Trade Comm'n, Data Brokers: A Call for Transparency and Accountability (2014), at vii (Commissioner Wright agrees that Congress should consider legislation that would provide for consumer access to the information collected by data brokers. However, he does not believe that at this time there is enough evidence that the benefits to consumers of requiring data brokers to provide them with the ability to opt out of the sharing of all consumer information for marketing purposes outweighs the costs of imposing such a requirement.).

Chart 7: Disclosed Use of Data

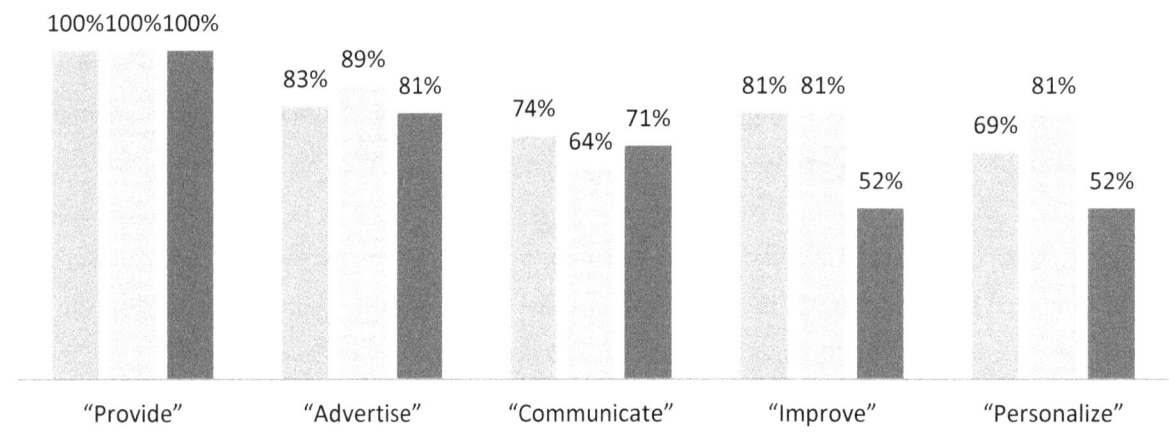

Price Comparison   Deals   ■ In-Store Purchase

Sample Size: Price Comparison: n = 42 | Deals: n = 47 | In-Store Purchase: n = 42

DATA SHARING

As with data collection and use, most of the apps' privacy policies had language explaining how consumers' data might be shared. While many policies prefaced their sharing sections with a sentence stating that they did not "sell or share" users' personal information "except as described" in the policy, almost all of the policies then used vague language that reserved broad rights to share consumers' data. Overall, 29% (13) of the "Price Comparison" policies, 17% (8) of the "Deals" policies, and 33% (14) of the "In-Store Purchase" policies reserved the right to share users' personal data without restriction.

Chart 8: Data Sharing

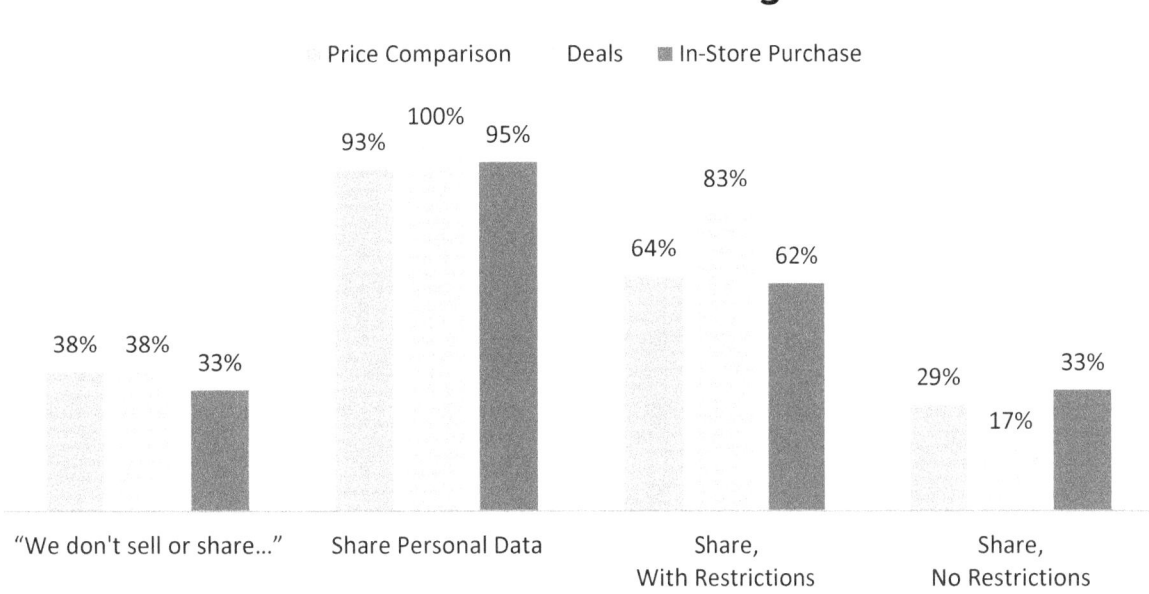

Price Comparison   Deals   ▓ In-Store Purchase

Sample Size: Price Comparison: n = 42 | Deals: n = 47 | In-Store Purchase: n = 42

The majority of the privacy policies, however, contained general language promising to restrict the sharing of users' personal data. As shown in Chart 9 below, a sizeable number of the policies stated that they would share users' personal data only with "service providers," who could use the data only to provide their services.[46] A few other policies specifically stated that recipients would be bound by confidentiality agreements. Still others provided that users' personal data would be shared only with a user's consent.[47]

Chart 9: Sharing Restrictions

Sample Size: Price Comparison: n = 42 | Deals: n = 47 | In-Store Purchase: n = 42

PRIVACY POLICY RECOMMENDATIONS

Consumers should be able to evaluate and compare the data practices of different services in order to make informed decisions about the apps they install. The number of readily available privacy policies addressing the collection, use, and sharing of data is a step in the right direction. However, many disclosures used vague language, reserving broad rights to collect, use, and share consumer information, rather than describe how the apps actually handle consumers' data.

---

46. Staff notes that there are likely to be a number of parties providing services within these apps, including advertising networks and analytics companies. Indeed, most of the apps' privacy policies reserved the right to use consumer data for advertising purposes. Staff therefore reiterates its recommendations that app developers, ad networks, analytics companies, and other relevant parties improve their disclosures. *See* Fed. Trade Comm'n , Mobile Privacy Disclosures, *supra* note 16, at 24.

47. In addition to the three categories of sharing restrictions described above and in Chart 9, staff also encountered several policies that contained partial restrictions, such as a promise to not share a consumer's name with third parties for the third parties' marketing purposes or a commitment to only share "non-sensitive personal information" with third parties.

Such disclosures preserve broad rights but fail to achieve what should be the central purpose of any privacy policy — making clear how data is collected, used, and shared.[48] Further, they suggest that these app developers may not be evaluating whether they have a business need for the data they are collecting. In the Commission's March 2012 Privacy Report, the FTC called for companies to build in privacy at every stage of product development, give consumers the ability to make decisions about their data at a relevant time and context, and make information collection and use practices transparent.[49] There, the FTC specifically called for app developers to limit their collection to the data needed for a requested service or transaction[50] and to make their privacy notices clearer, shorter, and more standardized to enable better comprehension and comparison.[51] While we are encouraged that many companies have privacy policies, the results of staff's survey indicate that app developers can do a better job of considering reasonable data collection and use limitations and describing those activities clearly to consumers.

SECURITY

Although smartphones are beginning to change the way people shop, a significant percentage of smartphone users have not used their mobile devices to make purchases.[52] Because consumers often identify concerns about security as a major impediment to their adoption of mobile payments technologies,[53] staff reviewed the policies for language related to security. Staff found that the overwhelming majority, over 80% of the reviewed policies, had language promising that the services took steps to ensure the security of consumers' data.[54]

Many policies assured users that the service implemented "technical," "organizational," and/or "physical" safeguards to ensure that consumer data was secure, and a number specifically provided that the service used encryption technology to transmit or store consumer data. In addition, many of the policies stated that they used "industry standard" or "reasonable" security technologies like "SSL" to protect consumer data, and some went so far as to promise that the service was "more secure than a bank," or "even safer than writing a check or using a credit card."

---

48. *See supra* note 44.

49. *See id.* at i.

50. *Id.* at 33 (Commissioner Ohlhausen does not support a strict data minimization requirement.).

51. *Id.* at 64.

52. One study, conducted in 2013, found that only "17 percent of all smartphone users have made a point-of-sale payment using their phone in the past 12 months." Consumers and Mobile Financial Services 2014 *supra* note 2, at 1.

53. *See id.* at 15-16 (reporting that 46 percent of respondents cited concerns about data security as a reason for not using mobile payments technology, making security consumers' "main reservation").

54. Specifically, 76% of the price comparison apps, 90% of the deals apps, and 90% of the in-store purchase apps promised to take steps to ensure the security of consumer data.

Chart 10: Policies with Specific Security Language

■ Price Comparison ■ Deals ■ In-Store Purchase

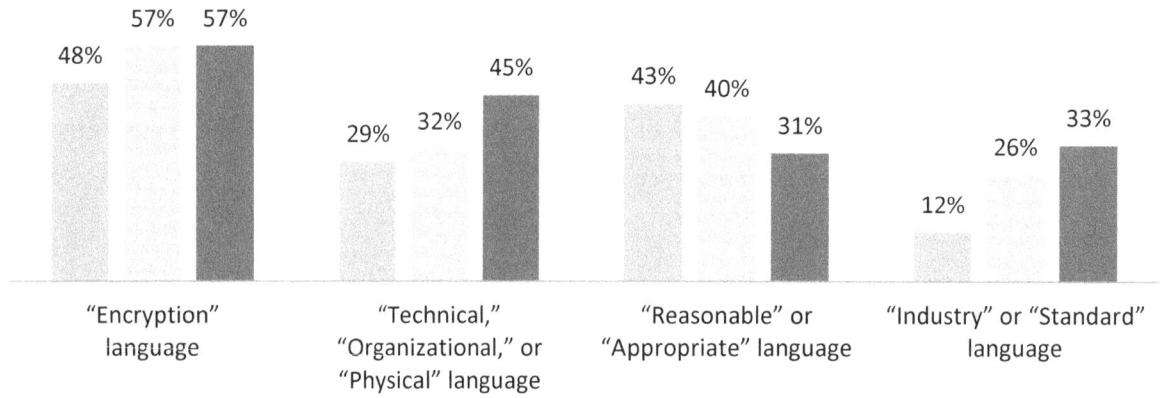

Sample Size: Price Comparison: n = 42 | Deals: n = 47 | In-Store Purchase: n = 42

As discussed during the FTC's April 2012 mobile payments workshop, technological improvements in today's smartphones offer the potential for increased data security for payment transactions.[55] Because staff did not test the services to verify the security promises made by these apps, staff cannot determine the level of security they provide. The FTC has addressed reasonable and appropriate security standards for mobile apps through both enforcement actions and business guidance materials.[56] Staff encourages vendors of shopping apps, and indeed vendors of all apps that collect consumer data, to secure the data they collect. Further, these apps must honor any representations about security that they make to consumers.

---

55. *See* Fed. Trade Comm'n, Paper, Plastic... or Mobile, *supra* note 9, at 11-12 (summarizing the technological advances — such as end-to-end encryption and dynamic data authentication — that were discussed by workshop panelists).

56. *See, e.g., Snapchat, Inc.,* FTC File No. 1323078 (F.T.C. May 8, 2014), *available at* http://www.ftc.gov/enforcement/cases-proceedings/132-3078/snapchat-inc-matter; *Credit Karma, Inc.,* FTC File No. 1323091 (F.T.C. Mar. 28, 2014), *available at* http://www.ftc.gov/enforcement/cases-proceedings/132-3091/credit-karma-inc; *Fandango, LLC,* FTC File No. 1323089 (F.T.C. Mar. 28, 2014), *available at* http://www.ftc.gov/enforcement/cases-proceedings/132-3089/fandango-llc; *Mobile App Developers: Start with Security,* Fed. Trade Comm'n (Feb. 2013), *available at* http://business.ftc.gov/documents/bus83-mobile-app-developers-start-security (guidance for app developers about how to incorporate robust security into mobile apps).

CONCLUSION

Whether using an app to find the best prices, redeem coupons and discounts, or pay for purchases, it is clear that mobile technology offers consumers new opportunities for in-store shopping experiences. However, staff's survey and report also reveal that these apps should provide more pre-download information about the consumer protections they provide.

Consumers should be able to make informed decisions about the apps they install and the services they use. In addition to being able to understand their rights and protections in case something goes wrong with a transaction, consumers should also be able to evaluate apps' data practices before signing up to use a particular service. Therefore, staff reiterates its recommendation that mobile payment companies provide clear dispute resolution and liability limits information to their customers. In particular, when prepaid funds are involved, consumers should be able to determine how a payment service works and what they can do if they encounter a problem. Further, companies providing mobile shopping apps should clearly describe how they collect, use, share, and secure consumers' personal and financial data.

Staff is committed to working with all stakeholders toward greater transparency and meaningful disclosure about consumers' rights and protections in the mobile marketplace. In the interim, before downloading an app, staff encourages consumers to learn how the service works, what they can do if they encounter a problem, and how their personal and financial data is collected, used, and shared. If consumers cannot find this information, they should consider taking steps to minimize their exposure by limiting the personal and financial data they provide, or by choosing a different app.

APPENDIX A

FTC staff examined shopping-related apps currently available for download in the two biggest app stores (Google Play and the iTunes App Store), focusing on three functionalities:

the ability to compare products and prices, in a store, using a mobile device;

the ability to obtain and redeem product or store discounts, using a mobile device; or

the ability to purchase a tangible product or service, in a store, using a mobile device.

This section provides additional information about staff's process for collecting and reviewing the data in the attached report.

Having a Pool of Relevant Apps

Before staff could begin its analysis, staff had to identify relevant apps and group them into manageable pools based on the three functionalities of interest. Staff achieved this in two steps. Staff first used a set of keyword searches and app store categories to find apps possessing functionalities of interest. Staff then reviewed the promotion pages of the returned apps in order to find the top 25 apps in each of the three functionalities of interest, in each app store.

FINDING RELEVANT APPS

Staff began its survey by identifying particular keyword searches and app store categories that could be used to discover relevant apps. In order to determine where and how apps purporting to offer relevant functionality could be found, staff ran a number of preliminary keyword searches and explored the apps found in various app store categories. Based on these exploratory searches, staff compiled a list of keyword searches and app store categories that returned the highest concentration of relevant apps. Those terms and categories are listed in Table A1 below.

Table A1

Mobile payments	Finance
Wallet	Lifestyle
Mobile pay	Shopping (Google Play); Catalogs (iTunes)
Price check	Productivity (Google Play); Utilities (iTunes)
Barcode scanner	Business
Coupon	
Point of sale	

Having identified relevant keyword searches and app store categories, staff next wrote a computer script capable of simultaneously running each keyword search and visiting each category in the Google Play and iTunes app stores. For each app returned by a particular search or category, the script recorded the app's name, developer, lower bound of the download range (for apps in Google Play), number of "all version" user feedbacks (for apps in iTunes), and the app promotion page's URL.[57] FTC staff applied the script to Google Play on May 30, 2013, which yielded 5,088 unique apps. Staff then removed any apps that had fewer than 5,000 downloads, resulting in a total first round pool of 3,062 unique Google Play apps. A week later, on June 10, 2013, FTC staff applied the script to the iTunes App Store, where it yielded 4,848 unique apps. Staff then removed those apps with fewer than 10 total user feedbacks, resulting in a total first round pool of 3,294 unique iTunes apps.

DISTINGUISHING RELEVANT APPS FROM IRRELEVANT APPS

Because the first round pools contained many irrelevant results, staff reviewed each app's promotion page to separate those apps that appeared to offer at least one of the three functionalities of interest from those that did not. Reviewers clicked on the live link to each app's promotion page, reviewed the pages in their entirety, and categorized the applications as price comparison apps, deal apps, and/or in-store purchase apps, or not relevant, according to the following criteria:

- **Price comparison apps** — To be included in this category, an app's promotion page needed to contain language mentioning barcode scanning, price checking or matching, comparing prices, or finding better deals nearby. Apps with promotion pages featuring screenshots that depicted barcodes, price lists, or storefronts suggesting that the app would enable a user to engage in comparison shopping were also included.

- **Deal apps** — To be included in this category, an app's promotion page needed to contain language mentioning coupons, deals at stores or restaurants, scanning or storing discounts in one place, or allowing the user to obtain special offers or discounts through some triggering activity.

- **In-store purchase apps** — To be included in this category, an app's promotion page needed to contain language such as "payments," "paying," or "purchase," in conjunction with specific stores, vendors, restaurants, or other retailers that appeared to accept the app in-store. Apps with promotion pages featuring screenshots that depicted point-of-sale terminals, cash registers, storefronts, or other images suggesting that the app would enable a user to make purchases were also included.

---

57. The script recorded this information for the first 400 results returned by each keyword search and for the top 300 "free" apps and the top 300 "paid" apps in each of the relevant app store categories. It also removed duplicate entries, recording the results in a single database.

Staff completed this phase of the review on June 17, 2013, the results of which are listed in Table A2 below.

Table A2

Price Comparison	151	99
Deals	240	236
In-store Purchase	34	26
Not relevant	2,672	2,930

Next, staff ranked the apps based on their relative number of installs. Google Play provides a download range for each app, so staff was able to rank the apps from most installed to least installed without making any assumptions. The iTunes App Store, however, does not provide any download information. In order to rank the iTunes apps by installs, staff relied on the number of user feedbacks (which iTunes does provide) as a proxy for number of installations. The top 25 apps in each functionality category for each app store were included in the survey.

Having established a manageable pool of apps centered on each of the three functionalities of interest, staff created sets of questions designed to measure the pre-download information available to consumers. Staff was interested in learning about each service's general behavior (how the services functioned), liability limits and dispute resolution procedures (what users could do in case something went wrong), and privacy and security practices (what information was collected, and how was it used, shared, stored, and transmitted). Staff then used generic Windows-based desktop computers to visit the app promotion pages and developer websites for each app, collecting information relevant to the services' general behavior, dispute resolution and liability limits policies, and privacy and security practices.

Staff's review included any license agreement, privacy policy, terms document, or other developer-provided disclosure that appeared to be associated with each service. Where a document was responsive to a question, staff preserved the document and quoted the relevant language in answering the question rather than characterizing the developers' practices. Staff began collecting the pre-download information on July 1, 2013, and finished on September 20, 2013. Once all of the initial data collection was complete, staff began analyzing the specific pre-download language that the developers provided.

Based on this detailed review, staff discovered that some services did not appear to have a U.S. presence or were no longer available. Eliminating these apps from the pools left 47 price comparison apps, 50 deal apps, and 45 in-store purchase apps. In total, staff reviewed the pre-download information associated with 121 unique apps.[58] As noted in the report, seven of the in-store purchase apps appeared to lack the ability to make purchases, and eight did not provide enough detail for staff to confirm that they contained such functionality.[59] Staff excluded these 15 apps from the report's dispute resolution findings but included them in the privacy and security review (since they raise many of the same privacy concerns).

Finally, because the attached report presents detailed findings regarding the content of the apps' privacy policies, staff explains how these findings were derived. In order to measure the content of the apps' privacy policies, staff first read each policy in its entirety, pulling out language that related to one of three questions: what data could be collected; how data could be used; and with whom data could be shared. Next, staff reviewed the specific language related to each of the three questions in order to measure the data points, uses, recipients, and restrictions specifically identified in the policies. Retaining the policies' language, staff recorded each data point, use, recipient, or restriction found in the policy.

Implications of Methodology and Interpretation of Results

Survey designs inevitably make trade-offs. Presenting evidence in ways that shed the most light on some questions necessarily leave other issues unaddressed. Staff's process for finding and analyzing mobile commerce apps permitted a comprehensive review of the information available to consumers regarding some of the most-installed shopping apps available today, but it is important to bear in mind several implications of the survey design.

The sample of apps in staff's review is not a perfect representation of mobile commerce apps. By focusing on three different functionalities of interest — price comparison, deals, and in-store purchasing — the survey ignores other commerce-related functionalities. This would include apps that focused on online commerce, person-to-person payments, or the purchase of virtual goods. Further, by focusing on the top 25 apps in each of the three functionalities of interest, in each app store, the survey leaves out a number of less popular but otherwise relevant services.

In addition, because staff was interested in measuring what a consumer could learn about a particular app before making the decision to use it, staff's review centered on pre-download information. It is possible that the apps provide their users with additional disclosures and information as the user interacts with the service, which may affect the privacy and dispute resolution and liability limits findings contained in the attached report. However, staff took several steps to minimize the effects of this omission on the report's findings. In order to be certain that staff's findings were as accurate as possible, staff downloaded each app and reviewed the information presented to a first-time user. Staff limited this portion of its review to information gathering, *i.e.*, staff did not create any accounts or otherwise use the apps. While many apps provided

---

58. This figure is less than the sum of the unique number of apps in each pool (142) because several apps were found in more than one pool.

59. *See supra* notes 23 and 32.

information upon launch, no app provided new information. That is, no app provided information that was not already available elsewhere on the developer's website or the app's promotion page. Second, staff limited its review of dispute resolution and liability limits policies to include only the 30 in-store purchase apps that clearly enabled in-store purchase functionality.

www.ingramcontent.com/pod-product-compliance
Lightning Source LLC
Chambersburg PA
CBHW080621180526
45168CB00007B/3010